HIBERNATION

First published in the United States
in 1991 by
Gloucester Press
387 Park Avenue South
New York, NY 10016

Library of Congress Cataloging-in-Publication Data

Stidworthy, John, 1943-
Hibernation / by John Stidworthy.
p. cm. -- (Animal behavior)
Includes index.
Summary: Explores the practice of hibernation as a means for winter survival, discussing the habits of a variety of animals such as hummingbirds, bats, chipmunks, and rattlesnakes.
ISBN 0-531-17309-7
1. Hibernation--Juvenile literature. [1. Hibernation.]
I. Title. II. Series: Animal behavior (New York, N.Y.)
QL755.S75 1991
591.54'3--dc20 91-2674 CIP AC

Printed in Belgium

The author, John Stidworthy, worked as a lecturer at London Zoo and the Natural History Museum in London. He is now a full time writer specializing in natural history.

The consultant, Steve Parker, has written more than 50 books for children on science and nature.

Design: David West
Children's Book Design
Designer: John Kelly
Editor: Jen Green
Consultant: Steve Parker
Picture researcher: Emma Krikler
Illustrator: Karen Johnson

Photocredits:
Cover and pages 5, 7, 11 right, 12 left and right, 14, 15 left and right, 18, 19, 23, 24 left and right, 26 left and 29 top: Bruce Coleman Limited; page 6 left: Marie-Helene Bradley; pages 6 right and 22 bottom: Planet Earth Pictures; pages 8 left, 10, 11 left, 20, 21, 25, 26 right, 27, 28 and 29 bottom: Oxford Scientific Films; pages 8 right, 13 and 22 top: J. Allan Cash Photo Library.

ANIMAL BEHAVIOR

HIBERNATION

JOHN STIDWORTHY

GLOUCESTER PRESS
New York · London · Toronto · Sydney

CONTENTS

INTRODUCTION

Winter is a hard time for animals. Many cannot survive this harsh season. Some animals live through it by being active and searching for food. Others survive by going into a special deep sleep, called hibernation.

Animals hibernate to survive the cold of winter and to conserve as much energy as possible. When an animal hibernates, body processes like heartbeat and breathing slow down. The animal may appear to be dead; in fact, it is in a very deep sleep.

Other species of animals survive the cold in lighter forms of winter sleep. Still other kinds of animals have developed very different strategies for surviving difficult weather conditions and extremes of temperature.

Most true hibernators are small **mammals** (warm-blooded animals with fur that feed their young on milk). A few species of birds also hibernate.

Woodchucks, or groundhogs, and other marmots are among the largest animals that hibernate. These animals are a familiar sight in summer, but disappear from view for winter. In some species the time of going into hibernation, and of emerging again in the spring, is very consistent. It can be predicted within a day or two.

Woodchucks emerge from hibernation at the same time in early February every year.

CONSERVING ENERGY

All animals need energy from food to survive, but they use up energy searching for things to eat. When food is scarce, or when an animal is not equipped to look for it, it makes sense to conserve energy by resting. Sleep and hibernation are both ways of conserving energy.

Waking and sleeping

For humans, day is the time to be active. We see well in daylight, and can go about our tasks. When night comes, we cannot see to do much. It makes sense to rest and save energy at this time. We lie down and go to sleep. Being active in daylight and inactive during darkness is known as being diurnal. Many animals are diurnal like us.

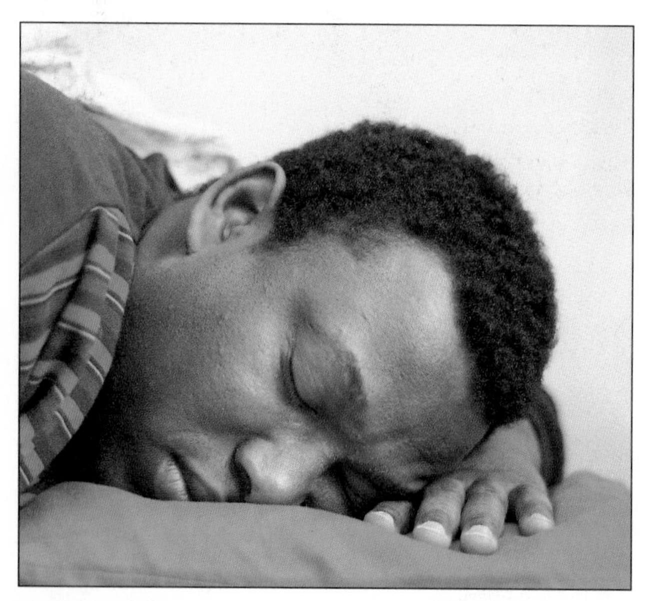

Humans are awake and active by day, and normally sleep when it is dark.

Animals like bats and moths are equipped to be active at night. They often spend the day asleep. These animals are called nocturnal.

Some animals take sleep one stage further each day or night. They "switch off" and their bodies slow down. They go into a kind of deeply sleepy state known as **torpor**. Some birds are torpid each night, and some kinds of bats are torpid during the day.

Active in winter

Just as there are many ways to survive the day or night, so animals have many methods of coping with the winter cold. Not all hibernate. Mammals such as badgers are still able to be active even in the depths of winter. But in cold weather, they spend more time in the warmth and comparative safety of their burrows. On really bad days they do not venture out. When food is scarce, they would use up more energy searching for it than they would get from it once it had been eaten and digested.

A badger looks out of its burrow, or sett. If the weather is bad in winter, it does not go out.

True hibernators

At the other end of the scale from animals that can be active in winter, are those creatures that sleep throughout the cold period, staying in a state of torpor for several months. Animals such as jumping mice, ground squirrels, and bats survive in this way. They do not wake up, even if there is a mild spell in the weather. These animals go into a deep "winter sleep" and are said to **hibernate**. This word comes from the Latin word which means winter.

In between creatures that are active all winter and those that hibernate, are animals such as hamsters. They spend much of the winter torpid, but wake from time to time to feed.

Aestivation

In some parts of the world, the worst weather is in summer, when it is uncomfortably hot and dry. Some animals react by finding a hiding place and becoming torpid. This "summer sleep" is called **aestivation**, from the Latin word for summer, and the animals that do it are called **aestivators**.

Baby swifts in the nest become torpid between feeds in cold weather, to save energy.

Tortoises hibernate in cooler climates as the fall turns cold. Pet tortoises should be put into a cool, frost-free place for hibernation. A box full of straw placed in an attic will keep a tortoise safe until warmth returns in spring.

SURVIVAL TACTICS

Animals hibernate to avoid periods of harsh weather, when it would be almost impossible for them to live normally. Hibernation is one way an animal can minimize its energy needs and avoid danger.

Cold-blooded animals
Most animals are cold-blooded. This does not necessarily mean they are cold, but that they make little heat in their bodies. They are usually about as warm as their surroundings. In a warm place they are warm; their bodies work well and they can be active. The colder they get, the less active they can be.

Some cold-blooded animals can partly control their temperature by their movements and behavior. For example, snakes and lizards bask in the sun to warm themselves, so that they can move around and hunt for the food they need for energy. If they become too hot, however, such animals will retreat into the shade to cool off.

In Europe's far north, adders hibernate for much of the year, and are active only in summer.

Honey bees are cold-blooded. They must beat their wings to warm themselves in the nest on a cool day.

Winter is a very difficult time for cold-blooded animals. When conditions are very cold for months on end, most of them become too cold to move. Creatures such as snakes and snails go below ground, or hide somewhere where they will not freeze. They stay there until warm weather returns.

Warm-blooded animals

Birds and mammals are warm-blooded. They can keep themselves warm and active even in cold surroundings, and be ready to hunt food or escape from danger. But all this activity means that these animals need a regular supply of food, to provide energy. They also need a good coat of fur or feathers to keep heat in. A polar bear, for example, has a long, thick coat of fur for insulation, and can be out and about, hunting seals, in the Arctic winter. Inside the coat, its body is kept as warm as it is in summer.

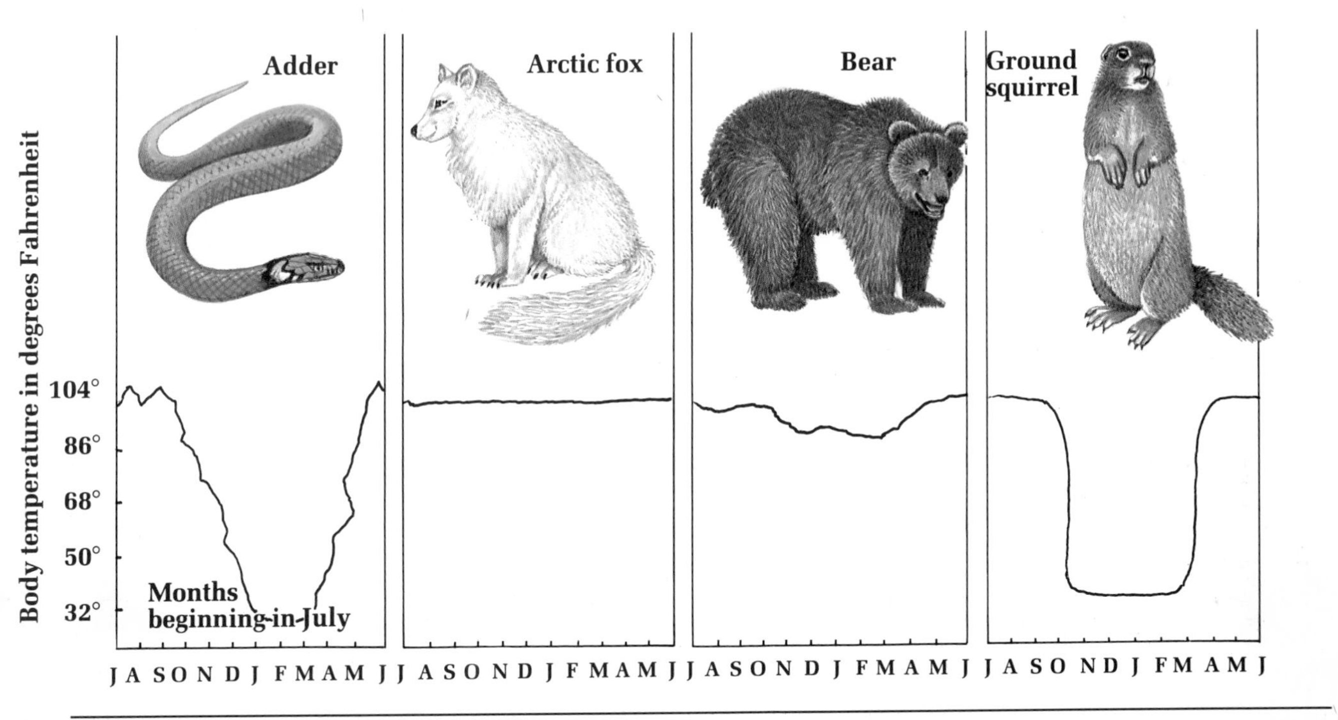

Long fur, furry paws, short, rounded ears, and a bushy tail all help the Arctic fox keep warm in icy places. The fox winds its tail around itself rather like a scarf when it is asleep.

The fur of an Arctic fox is up to 6 inches long in places, enabling it to survive temperatures of −40°F.

Smaller mammals have more difficulty keeping warm. They would be unable to move in fur coats as long as the Arctic fox's, so their insulation cannot be as good. Another problem is that a small animal has a large surface area of skin compared to its body volume, so loses heat more quickly. Some smaller creatures survive by burrowing to escape the worst of the cold. Lemmings run through tunnels under snow in winter, and are not exposed to even lower temperatures above ground.

But not all small mammals can burrow in this way. For these animals, the answer is to hibernate. With body temperature and activity cut down, the need for food is reduced to a minimum.

HOW ANIMALS HIBERNATE

Hibernation has many similarities to normal sleep. But in hibernation there are greater changes to body processes than in sleep. There are relatively few kinds of animals that can truly hibernate.

A lion is one of the few animals that can afford to be totally relaxed when resting. This lioness, sleeping in sunny, warm conditions, stretches out her legs and body and sinks into a deep sleep. But although relaxed, if she is disturbed she can be instantly alert.

In sleep and in hibernation
When animals sleep, they usually go to a safe place, protected from their enemies. Often, they sleep in a special posture. They may stretch out flat, or curl into a ball. Breathing and heartbeat are slower than when they are active. Body temperature may be slightly lower than in busy periods. But even in sleep, animals are aware of their surroundings to some extent. If there is a loud noise or sudden jolt, a sleeping animal wakes up.

When a creature hibernates, it also hides away in a nest or burrow, and takes up a sleep posture. But in hibernation, an animal becomes mostly unaware of its surroundings, and is almost impossible to rouse.

Saving energy

During hibernation, many of an animal's body functions slow right down. It may be scarcely breathing. Its heartbeat becomes slow and weak, and it makes hardly any urine (liquid wastes). Its temperature drops many degrees. In some hibernators, body temperature is only just above that of the surroundings. Some rodents may have body temperatures of only 35°F, just above freezing, instead of the more usual 98°F. Maintaining a low body temperature saves energy. A hibernator's use of energy may be 70 to 100 times less when hibernating than when active.

The dormouse's body is "switched off" during hibernation. It uses enough energy to survive.

Yet a hibernating animal is not completely out of action. If surrounding temperatures plunge dangerously low, a hibernating mammal can automatically switch on heat production in its body to keep itself from freezing. If they get too cold, bats hibernating in a cave can wake up and move to a different part of the cave to be in the best hibernating temperature. Marmots might wake up every month or so to urinate.

Antifreeze

Most cold-blooded animals spend the winter hidden in a place with a constant cool temperature. Frogs, terrapins, and some fish find refuge in the mud at the bottom of lakes. Here they are protected from freezing, but their bodies are cool and use little energy.

Like other terrapins, a diamondback spends the winter buried in the mud at the bottom of a lake.

Some cold-blooded animals spend the winter in exposed places. Some kinds of butterfly overwinter either as adults or pupae on the stems of plants, out in the open. These insects make chemicals in their body fluids which act as antifreeze. Some use just the same chemical that we put into car radiators.

The hedgehog cannot find insects and slugs to eat in winter. It saves energy by hibernating, often in a pile of leaves. But it might be active briefly if winter weather is mild.

THE RHYTHM OF LIFE

Animal behavior occurs in patterns, or rhythms. These patterns coincide with days, months, or seasons, or with tides. Hibernation is one of the ways a seasonal rhythm can affect some animals' lives.

Human bodies run to a daily rhythm. We sleep at night, even if we have done little to tire ourselves during the day. Bats also follow a 24-hour cycle, called a circadian rhythm, but they sleep by day and wake at night.

On the seashore, animals such as crabs have periods of activity and rest in time with the tides. They seem to have internal "clocks" that maintain this tidal rhythm, even if they are moved to a place where there are no tides.

Anemones and periwinkles close up while the tide is out, but become active when the tide returns.

A robin sings from its perch. In spring robins come into breeding condition, and the males start to challenge others for possession of a breeding area, or "territory." Their songs warn other males away.

Animal "clocks"

Some animals are geared to a monthly rhythm, their clocks keeping time with the moon month of 29 days. Yet others have annual rhythms, lasting one year.

Annual rhythms help bring about events such as nesting in birds. This is especially true in long-distance migrants, such as willow warblers. A particular change in the outside world often sets the animal's clock accurately. Increasing amounts of daylight in spring help to set a bird's breeding clock, and make its reproductive system ready to produce eggs and breed.

Time to sleep

Many hibernators have very accurate annual clocks. They start to hibernate at the same time each year, regardless of the weather at that particular time. For example, the golden-mantled ground squirrel of western North America always starts its hibernation in October.

Other hibernators need the stimulus of a cold spell to send them into hibernation, but even these have a seasonal rhythm. During the summer, they have a completely different reaction to cold weather. When exposed to cold, their bodies try to produce extra heat to keep them warm.

Desert heat

Rhythms control animals that aestivate, too. The large-toothed suslik, which lives in Central Asia, always starts its summer sleep in the first half of June. This is the time when plants start to wither in the summer heat.

The golden hamster prepares itself for hibernation, carrying food to its store in its cheek pouches.

In desert areas the most difficult weather for animals is often the heat of summer. The body rhythms of a large-toothed suslik cause it to aestivate in the high summer. It goes below ground before the heat becomes too intense, and the lack of food too acute.

FATTENING UP

Many animals prepare for winter by putting on fat, which provides body insulation and acts as an energy store. This is especially true of hibernators. Another tactic is to gather food and store it, so it is available for snacks in the middle of winter.

As winter approaches, the amount of daylight in each 24-hour period lessens. This is a signal to some animals to store food. Tree squirrels, which can be active in winter, bury nuts and seeds. On winter days when they venture from their nests, they can dig up a ready supply. They can detect pine seeds or acorns buried 12 inches below ground, although the seeds that they dig up might not be the ones that they themselves buried!

Some creatures that have a deeper winter sleep also store food, and may awake periodically to nibble it. Hamsters transport food in their cheek pouches to a chamber in their burrows.

Animals can also store a "food supply" in their own bodies. They can put on fat. For some animals that are active in winter, only a small amount of fat is needed, to help with insulation and provide an energy reserve for days when food is scarce.

The red squirrel takes advantage of the abundance of nuts in the fall to lay in a store for winter days.

Hibernators such as ground squirrels, and aestivators such as susliks make burrow systems with special chambers for particular activities. Some areas may be used as sleeping chambers, others are made into foodstores.

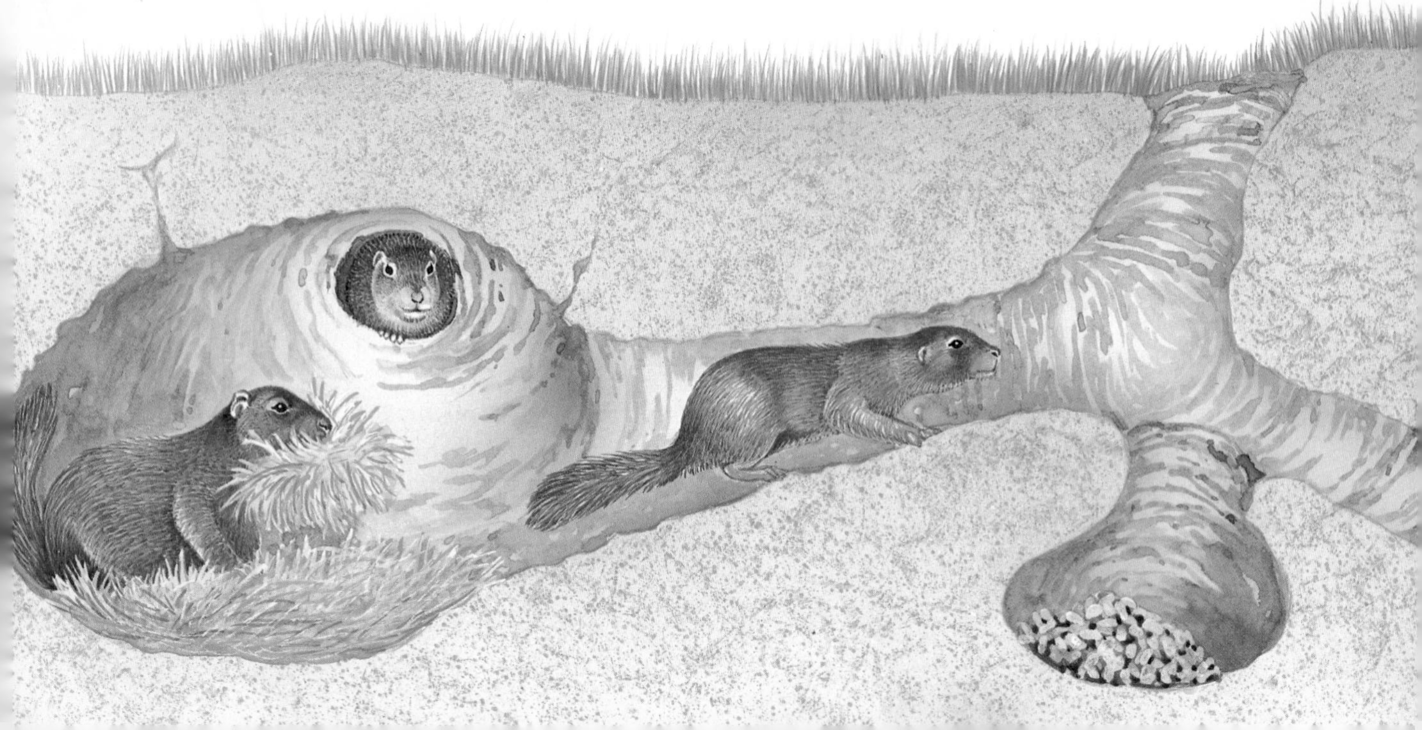

The grizzly bear gorges itself on the autumn crop of berries to put on a layer of fat for winter.

Putting on weight

True hibernators must build up much more fat. Changes take place in these animals' brains so that they never feel satisfied with a meal, and are always eating and nibbling at food. They move around less, and put on fat quickly. During the summer a woodchuck's body is less than one-twentieth fat. By the time it hibernates, fat may be one-seventh of its body weight.

Dormice (their name means "sleepy mice") spend half or more of their lives in hibernation. They prepare by feeding well in the fall, and become very fat. Half of the weight of a European dormouse is fat when it is ready for hibernation.

In the fall the edible dormouse needs to put on fat to hibernate. It feeds on fruit, nuts, some bark, and also insects.

The ancient Romans made use of the prehibernation habits of edible dormice by overfeeding them in special containers until they were very plump, and then feasting on them.

In Africa, a similar fate befalls fat mice today. These rodents are aestivators and store fat to tide them through the summer drought. But this puts them at risk of being hunted and eaten.

Brown fat

Much of the fat in a hibernating animal is a special type known as brown fat. This can be "burned" in the body to give a very rapid temperature rise when needed. This provides a safety net if the surroundings become exceptionally cold, and the animal is in danger of freezing to death. A hibernating ground squirrel is able to heat itself up by 55°F in three hours. The brown fat is also burned when the animal needs to wake up in spring.

A PLACE TO HIBERNATE

In autumn, hibernators prepare for the winter by finding a suitable sleeping place. It must be safe from enemies and protected from the weather. It may be a burrow, a specially made nest, or even a cave.

In winter, few small animals are out and about. Unseen, they struggle to survive winter, in various ways.

Some species of insects die off in the fall but leave a few adults to overwinter and breed next year. The common wasp does this. The workers die, but some queen wasps survive the winter, sheltering in woodpiles, in the slats of fences, or in garden sheds. Slugs and worms go deep underground to avoid frost. Snails shelter in crevices in old walls. Frogs and crayfish bury themselves in the mud at the bottom of ponds and lakes during the winter.

A queen common wasp emerges in spring, having survived the winter in a woodpile.

In its nest, a bundle of grasses and bark, a dormouse hibernates through the cold of winter.

Finding the right spot

Warm-blooded animals that hibernate must also choose a place to rest with care. Even though dormice build up plenty of body fat, probably fewer than half of them survive winter. The choice of a spot to hibernate can make all the difference. The dormouse's summer nest is a ball of grasses or shredded honeysuckle bark above ground. But for the winter it makes a much more protected nest, among tree roots or underground.

Jerboas of Central Asia dig complex burrows with several chambers for hibernation, some 12 inches below the desert surface. Marmots hibernate as a family, close together in an underground den. They block the entrance to their burrow with earth and grass.

Clustered together in large groups, pipistrelle bats hibernate to survive the winter. They choose places where conditions change very little. The presence of so many bats helps to conserve heat within the group.

A place to roost

Bats have to be particularly careful about where they hibernate. Their bodies are small but have a large surface area, and they will die if they lose too much water through their skins. To survive, they need to hibernate in damp places like caves, covered in droplets of water (condensation). A dry atmosphere can kill them very quickly.

Some kinds of bat, such as North American red bats, can survive a surrounding temperature as low as 32°F, and are able to hibernate in hollow trees. Greater horseshoe bats prefer a more constant temperature, around 50°F, for hibernation. Caves are ideal roosts for this species.

Some bats cluster in large numbers when hibernating. Each bat produces a tiny amount of heat, and so the whole group is kept from freezing. A few species, such as the American little brown bat, may form groups of thousands in hibernation.

JOURNEY TO SLEEP

Choosing the right place to rest or hibernate is the key to survival. Some animals need to make only a local search. Others migrate many miles to find the conditions they need to sleep.

Many hibernators find their winter shelter locally. Most small animals, after all, are unable to travel far. They must make the best of any crevices, underground chambers, or burrows that they can find or make.

Animals with wings can travel long distances more easily. Many birds fly to warmer climates for winter. Some insects and bats also fly dozens or even hundreds of miles to find the conditions they need for hibernation.

Monarch butterflies migrate in huge clouds, journeying thousands of miles to hibernate.

Incredible journey
One of the most remarkable animal migrations is that of the monarch butterfly of North America. Monarchs that live in the Great Lakes region fly south in the fall before the snows arrive. They journey 1,900 miles to Mexico. There they group on trees in a few locations, sometimes in clusters of many thousands together, and spend the winter resting. In spring they make the journey back. Many die on the way, but the migration probably results in more survivors than if the butterflies tried to live through the freezing conditions of the northern winter.

Many bats travel long distances to hibernate. Some bats, such as these little brown bats, follow regular migration routes to caves that are used year after year. Interference with traditional hibernating caves can be disastrous for some bat populations.

Many kinds of bats also make long journeys to find suitable hibernation sites. Their summer daytime roosts are in crevices in bark and similar hiding places. In winter these would be too cold and variable. A constant cool cave is required. To find a suitable one they must migrate.

Tracking migrations
Scientists have put wing clips on the forearms of little brown bats in North America to track their movements. In autumn the bats journey to particular caves to survive the winter in hibernation. In spring they return to their summer hunting grounds. Most travel 50 miles or less in each direction, but some fly nearly 180 miles to reach a suitable hibernation site.

Bogong moths of Australia make a long journey to aestivate. During November, the southern spring, flights of these moths can be seen moving up into the mountains. They spend the summer huddled together in caves and crevices in granite outcrops, resting. High in the mountains at 5,900 feet, they avoid the worst of the intense heat. One disadvantage of bunching togther, however, is that they make an easy target for predators, including humans. In February they make the return journey to the plains, to mate and lay eggs.

TRUE HIBERNATORS

Those animals that hibernate most completely have many special features in the way their bodies work and in their behavior, that make them different from other kinds of animals.

True hibernation is found only in certain small species of mammals, not in larger animals. These small mammals cool down rapidly, and can also warm up again quickly when necessary. Animals that have larger bodies are not able to do this.

True hibernators go into a very deep winter sleep. They are very still, and can be almost impossible to wake, even if moved. The heartbeat of a hibernator becomes very slow, and the animal breathes less often. Hibernating dormice may stop breathing for minutes at a time.

Body temperature

The body of a hibernator may feel very cold to our touch. Many regularly have body temperatures that drop to 36°F, just above freezing. In some animals, body temperature actually falls below freezing. The red bat can have a temperature as low as 23°F. Its blood and body fluids are kept from freezing by all the minerals and body salts in them. Such an animal seems more dead than alive. In fact, its body is using just enough energy to keep the animal alive and to stop it succumbing to the cold.

Flying squirrels have a large surface area of skin which loses heat quickly. Some species hibernate.

Stopping loss of water from the body during hibernation is also important. The brain switches off the production of urine by sending out a hormone, a special chemical messenger, around the body in the bloodstream.

The microscopic cells that make up the body seem to undergo a change in a hibernator. They still work, even if only very slowly, at temperatures lower than they could at other times of the year.

Out of action?

It would be wrong, however, to think that a hibernating animal's body is doing nothing. The brain is still supplied with blood, and is kept relatively warm. The brain's "thermostat," the part that regulates the body's temperature, is still working, though it is set to a lower temperature than normal.

Little is known about the body changes in many cold-blooded animals that spend the winter inactive. Some insects are known to produce "antifreeze" in their tissues. Others, including some moth larvae, are able to become deep frozen during winter and still thaw out alive the following year.

	Hibernating	Active
Heartbeat	1-10 beats per minute	100-200 beats per minute
Body temperature	35-50°F	95-104°F
Unconscious	Continuously	Sleeps by day
Water loss	Almost none	In feces and urine
Breathing rate per minute	May be less than 1	50-150

The table above shows some of the body functions of a true hibernator, such as a dormouse, while active and during hibernation.

The golden-mantled ground squirrel is a deep hibernator, disappearing into its burrow to sleep from autumn until March. It stores food underground. This store provides an immediate food supply when it awakens in spring.

RESTING IN WINTER

Many animals that do not truly hibernate spend more time in a shelter during winter, much of it asleep. Some make use of this time to rear their babies.

Skunks do not hibernate, but they spend a lot of time asleep in their dens during winter.

Animals that do not hibernate may still change their behavior patterns in winter. For example, tree squirrels remain in their nests in very cold weather, spending days at a time in these snug surroundings. Often they are asleep, but they show none of the special body reactions of a hibernator. In milder weather they venture out for an hour or so to forage.

Skunks, which are nocturnal animals, follow a similar pattern. On the worst winter nights they conserve energy and do not come out in search of food. Raccoons, too, spend most of the time in their dens in winter. Sometimes they stay there for a month at a time. But they are not hibernating. Their body temperature is normal and they are alert.

During the winter, seven-spot ladybugs are often found clustered together on plants. Two-spot ladybugs often overwinter in houses. Surprisingly, they can stand freezing conditions, but if kept too warm they will not survive until spring.

Close to hibernation

Certain animals have a winter sleep which is almost hibernation, though they are not true hibernators. Some bears, for example, spend half the year in their dens. Here they sleep and doze the winter away. Their body temperature is lower than when they are active, but only by about 9°F. Bears which are "cool" like this can still wake very quickly. But this slight reduction in body temperature cuts their energy needs by about half, so the fat stored in their bodies lasts longer.

The first months of a brown bear cub's life are spent in its mother's winter den, kept warm by the mother's body. A newborn bear can weigh less than a pound so the protection of the den is very important.

By the time polar bear cubs leave the den with their mother in spring, they are well grown.

Brown bears and black bears den up for the winter in cold areas, but polar bears are active through the winter. The exceptions are pregnant female polar bears, who dig dens in which to rest and have their babies. The cubs are born in the middle of winter, blind, helpless, and tiny compared to their mother. She suckles them on very rich milk, almost one-third fat. This helps give them the energy to keep warm as they grow, and by the spring the cubs are strong enough to follow their mother when she leaves the den. Brown and black bears also produce their cubs during the winter.

DAILY SLEEP

Animals do not waste energy. Sleep is an excellent means of conserving it, and many animals spend as much time sleeping as they can. Some small animals become torpid each day to save energy.

Sleep patterns

When their daily needs are satisfied, most animals tend to rest. Apes such as chimpanzees have a sleeping pattern rather like ours, spending each night dozing for six to eight hours. Some animals sleep much more. Lions catch nourishing food, and need to hunt only once or twice a week. They sleep on average more than 16 hours a day.

Animals that have low energy needs and live surrounded by food can sleep even longer. A sloth lives most of its life hanging from the branches of the tree whose leaves it eats. It does not need to move far for a meal, and its slow life does not require much food. It often sleeps 20 hours a day.

The three-toed sloth has a very slow lifestyle. Its body is not quite as warm as a human's.

Some large plant-eaters (herbivores) are at the other extreme. They need to eat a great deal to keep their large bodies going, and they are always watching out for enemies. They have little time for sleep. Horses sleep for only about five hours a day, often on their feet. Elephants may lie down to sleep for less than four hours in the night.

A giraffe spends most of the day feeding. It sleeps for just a few minutes at a time.

Daily torpor

Some animals go further than sleep each day. They become torpid. The tree dormouse, a nocturnal animal, does this. In its daytime state it may take a breath only every quarter of an hour. At night it becomes active again, and breathes dozens of times a minute.

Hummingbirds become torpid each night. Most are tiny creatures, and, like many birds, their body temperature is even higher than mammals, at about 104°F. Hummingbirds lose heat quickly in cool conditions. They also need energy for their fast-beating wings. In the day, they can see to find their food – nectar from flowers, and some insects.

A hummingbird probes a flower for nectar. A hummingbird's body uses up energy at an enormous rate. Hummingbirds feed to restore this energy by day. At night they cannot afford a huge energy loss, and become torpid.

Nectar is good fuel, but a hummingbird needs to feed every ten minutes. In one day it takes in almost its own weight of sugar. At night the surroundings are cooler, and the hummingbird cannot see to feed. So it becomes torpid. Its temperature drops as it perches on a twig. It stays rigid through the night, becoming active again in the morning.

Two horseshoe bats sleep in the daytime, until night returns and they become active.

Many species of bats also become torpid daily. When awake and active at night, they can have a body temperature higher than humans. Bats have been measured with a temperature of up to 108°F in flight, compared to 98°F for humans. Bats gather their insect food in a burst of nighttime activity. Then they settle to digest their meal with a slightly cooler body, at around 86°F. Finally, their temperature may drop to that of their surroundings as they take their daytime sleep. This rhythm minimizes the bat's energy needs, and is very useful to a small animal with an unpredictable food supply.

DRIED-UP SURVIVORS

Some mammals survive heat and drought by aestivation. Many cold-blooded animals can also go into a dormant state, and survive very dry conditions, and high or low temperatures.

Dwarf lemurs, bats, and some ground squirrels are all mammals that aestivate to avoid the driest weather of the year. Yet their survival abilities are exceeded by some cold-blooded survivors. Some of these creatures can stop functioning completely for months or years, showing no sign of life. Then they can be active once more when conditions become more favorable.

Dwarf lemurs in Madagascar avoid the dry season by becoming torpid in dens in hollow trees.

Becoming dormant

Many microscopic single-celled animals, such as amoebas, go into a dormant state when their surroundings dry up. These tiny creatures make a tough skin around their bodies, shut down their activities, and dry up. They seem dead, but when water reappears, sometimes years later, they can absorb it and return to their normal working state.

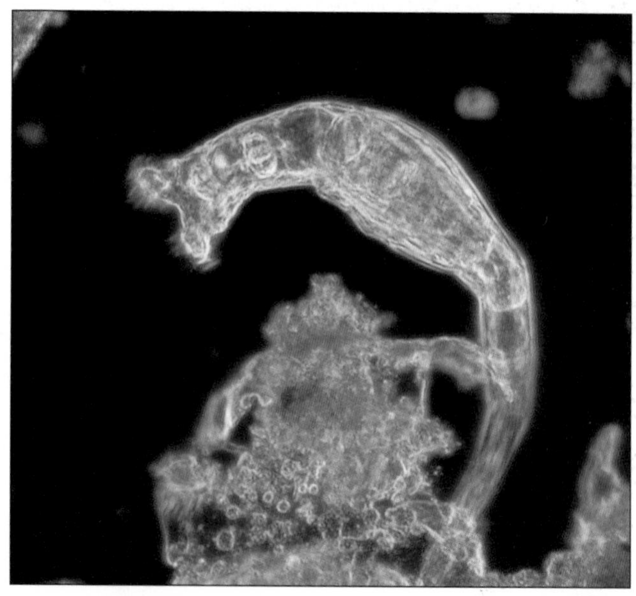

Rotifers like Philodina are great survivors, coming back to life after a long period of drought.

Rotifers are tiny, many-celled creatures that live in pond water and other damp places. They, too, can survive drought. One species has been known to survive for more than 100 years "freeze-dried" in Antarctica, and then to recover. Water bears are another type of microscopic animal that can become dormant and survive many years of dryness, and also temperatures far below freezing. They become active again when water returns.

An African lungfish burrows into the mud to avoid drought. In its hardened cocoon it is quite well protected. It leaves breathing tubes up to the surface so it can obtain the small amount of oxygen it needs in its dormant state.

Even some fish can endure drought. The North American bowfin becomes torpid when water is drying out, and is able to use its airbladder to breathe air. Lungfishes, too, can breathe air. African and South American species can survive in places where the pools of surface water dry up completely from time to time. They burrow into the still-damp mud and secrete a cocoon of slime which hardens around them. Inactive, breathing very little, and living off stored energy in their body tissues, they may survive for years in this state.

Damp-loving animals

Snails are usually thought of as animals of damp places. Yet garden snails can make a "door" of hardened slime to plug their shell openings. This helps them to endure dry weather. Some snails even manage to live in desert surroundings, although these might spend much time inactive. Worms, too, go deep underground and secrete a cocoon of slime around themselves as a protection against drought.

Garden snails can plug their shells to prevent water loss in dry conditions.

WAKING UP

In spring, an animal has to wake itself up from hibernation. It quickly needs to resume normal activities such as feeding. For many animals, spring is also the time for mating and rearing young.

In many kinds of animals, especially cold-blooded ones, the warmer temperatures of spring bring about the awakening from winter inactivity. Insects may be stimulated to hatch from their eggs. Increasing warmth and daylight each day stimulate activity and mating behavior in frogs and reptiles. It is important for reptiles to keep warm as their bodies prepare for this time of activity. In temperate climates, snakes such as adders and garter snakes bask in the sun in the early spring whenever they can, to absorb warmth.

North American garter snakes cluster together in the den for hibernation.

Male adders fight in spring, wrestling to see which is strongest and will mate with the females. Adders bask in the sun in early spring, so that their bodies are warmed and they can be active.

Returning to normal

When a true hibernator wakes up, all the changes which took place as it settled down for winter are reversed. For example, the heart of a dormouse speeds up again, and its body temperature returns to normal. To do this, the animal burns body fat. The amount of fat it has left at the end of winter is critical. It needs enough to warm itself and tide it over until it finds a new supply of food. If it has too little, it will die soon after it comes out of hibernation. To solve this problem, some hibernators store food to eat when they emerge in spring.

Secrets of winter sleep

There are still many mysteries about hibernation and the process of awakening. How do animals in constant conditions, in an underground den with hardly any change in temperature or light levels, know when it is time to emerge in spring? They may begin to come out before the change in the weather could really have affected conditions deep underground. How do snakes, for example, know that it is sunny? How do they get warm enough to move in the first place?

Flying squirrels hibernate in winter and emerge when warm weather returns in spring.

Why do woodchucks, deep in their tunnels, always emerge from hibernation about the same week at the beginning of February, year after year? Scientists are carrying out research to try to answer these puzzles, but at the present time, many questions about hibernation remain.

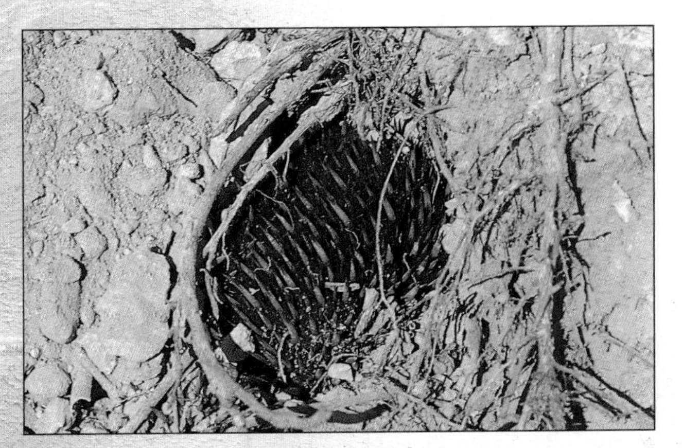

Echidnas are strange egg-laying mammals, whose periods of activity are linked with the weather.

SPOT IT YOURSELF

You can study animals and the way they behave almost anywhere. Learn to detect animals by the signs they leave: burrow entrances, nests, footprints in mud or snow, hair caught in wire or branches, droppings, half-eaten leaves, and discarded shells. When nature-spotting, keep as still and quiet as possible. Keep a record of when hibernators disappear in winter, and when they emerge again.

Practical tips for nature-spotting
Wear wind- and waterproof clothing in dull colors. Polaroid glasses reduce surface reflection for seeing underwater. A lens magnifies small animals and a camping mat gives some comfort.

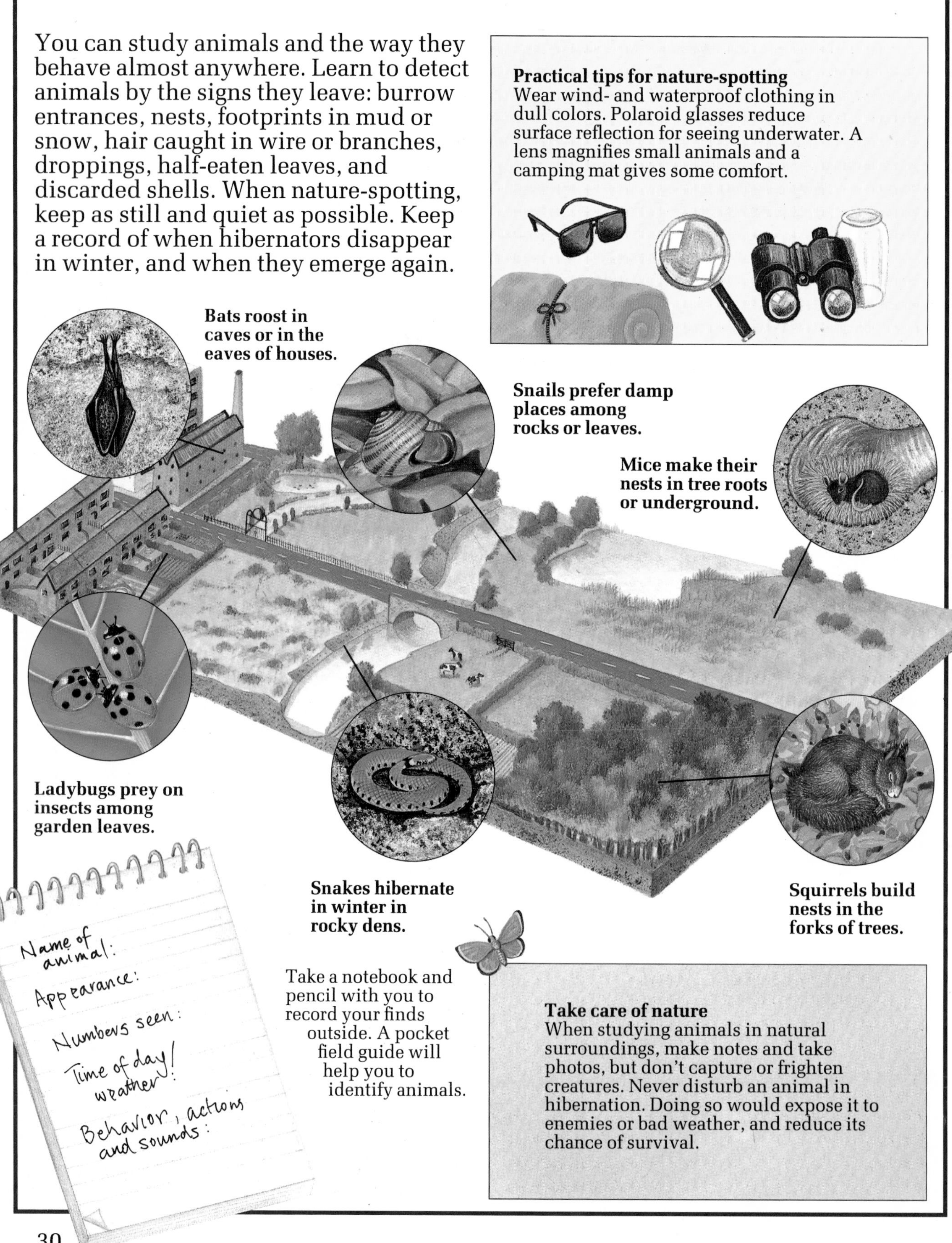

Bats roost in caves or in the eaves of houses.

Snails prefer damp places among rocks or leaves.

Mice make their nests in tree roots or underground.

Ladybugs prey on insects among garden leaves.

Snakes hibernate in winter in rocky dens.

Squirrels build nests in the forks of trees.

Take a notebook and pencil with you to record your finds outside. A pocket field guide will help you to identify animals.

Take care of nature
When studying animals in natural surroundings, make notes and take photos, but don't capture or frighten creatures. Never disturb an animal in hibernation. Doing so would expose it to enemies or bad weather, and reduce its chance of survival.

GLOSSARY

aestivate to go into a state of torpor in summer.

aestivation summer torpor.

annual yearly, each year, at yearly intervals.

brown fat a special type of fat found in hibernating animals. Hibernators burn brown fat to warm themselves up quickly.

circadian rhythm a rhythm with a period of about one day.

cocoon a shell or covering an animal makes around itself. Animals may stay in a cocoon while they are dormant.

cold-blooded animal one that is unable to make heat to warm its body, and that is therefore dependent on heat from its surroundings.

diurnal active in the daytime.

dormant in an inactive state.

hibernate to become torpid or dormant during the winter months.

hibernation winter torpor.

hibernator an animal that hibernates.

hormone a chemical that travels round an animal's body in the blood and acts as a messenger between parts of the body.

insulating keeping heat in or out.

insulation a body covering that keeps heat in or out.

mammal an animal that is warm-blooded and feeds its young on milk.

migrant an animal that migrates.

migrate to move seasonally from one area to another.

migration a seasonal movement of animals.

minimum the smallest possible.

nectar a sugary fluid made by flowers.

nocturnal active at night.

posture the particular way an animal stands, sits, or lies down.

pupa "resting" stage of insects such as beetles, bees, and moths, between the larva stage and the adult.

rodent a type of plant-eating mammal, for example a rat or squirrel, with large front teeth.

stimulate to cause an animal to do something.

strategy a way of doing something that may help an animal to survive.

thermostat a device, or part of the body, that helps to maintain a constant temperature.

torpor a deeply inactive state, in which the body functions slow down.

warm-blooded animal one that can maintain a constant body temperature, whatever the temperature around it.

INDEX